Generis
PUBLISHING

Effects of Symmetry-Breaking on the Dynamics of the Shinriki's Oscillator

Léandre Kamdjeu Kengne

Title: ***Effects of Symmetry-Breaking on the Dynamics of the Shinriki's Oscillator***

ISBN: 978-1-63902-839-9

Author: Léandre Kamdjeu Kengne

Cover image: Author

Publisher: Generis Publishing
Online orders: www.generis-publishing.com
Contact email: info@generis-publishing.com

Effects of Symmetry-Breaking on the Dynamics of the Shinriki's Oscillator

Léandre Kamdjeu Kengne [a,,1,2], Romanic Kengne [1], Zeric Tabekoueng Njitacke [3], Theophile Fozin Fonzin [4], Justin Roger Mboupda Pone [2], Hervé Thierry Kamdem Tagne [5]

[1] Unité de Recherche de Matière Condensée, d'Electronique et de Traitement du Signal (UR-MACETS), Department of Physics, University of Dschang, P.O. Box 67, Dschang, Cameroon

[2] Unité de Recherche d'Automatique et Informatique Appliquée (UR-AIA), Department of Electrical Engineering, IUT-FV Bandjoun, P.O. Box 134, Bandjoun, Cameroon

[3] Department of Electrical and Electronic Engineering, College of Technology (COT), University of Buea, P.O. Box 63, Buea, Cameroon

[4] Department of Electrical and Electronic Engineering, Faculty of Engineering and Technology (FET), University of Buea, P.O. Box 63, Buea, Cameroon

[5] Unité de Recherche de Mécanique et de Modélisation des Systèmes Physiques (UR-2MSP),

Department of Physics, University of Dschang, P.O. Box 67, Dschang, Cameroon

[a] Corresponding author: Léandre Kamdjeu Kengne

[a] Email address: lkamdjeu@gmail.com

Abstract

We investigate the dynamics of the well-known Shinriki's oscillator both in its symmetric and asymmetric modes of operation. Instead of using the classical approximate cubic model of the Shinriki's oscillator, we propose a close form model by exploiting the Shockley exponential diode equation. The proposed model takes into account the intrinsic characteristics of semiconductor diodes forming the nonlinear part (i.e. the positive conductance). We address the realistic issue of symmetry breaking by considering different numbers of diodes within the two branches of the positive conductance. The dynamics of the system is investigated by exploiting conventional nonlinear analysis tools such as bifurcation diagrams, phase space trajectories plots, basins of attractions, and graphs of Lyapunov exponent as well. In the symmetric mode of operation, the system experiences coexisting symmetric attractors, period doubling route to chaos, and merging crisis. The symmetry breaking analysis yields two asymmetric coexisting bifurcation branches (i.e. asymmetric bi-stability) each of which exhibits its own sequence of bifurcations to chaos when monitoring the main control parameter. In this special mode, the merging process never occurs; instead, one of the bifurcation branches vanishes when decreasing the control parameter beyond a critical value following a crisis event. The theoretical results are validated by carrying out laboratory experimental studies of the physical circuit.

Keywords: Shinriki's oscillator; symmetry breaking; mathematical modeling; coexisting bifurcation; basins of attraction; experimental study.

TABLE OF CONTENTS

1. Introduction

Symmetric structures are common in nonlinear circuits generating chaotic signal. Examples include Chua circuit [1], autonomous Duffing van der Pol oscillator [2], Shinriki's circuit [3,4], jerk circuits [5,6], and the autonomous Silva-Young oscillator [7], just to cite a few. Connecting in antiparallel a semiconductor diodes pair represents one of the simplest design strategies to obtain a symmetric structure/component [5-7]. The evolution equations of these circuits are transformed into themselves as a result of symmetry operations, for instance an inversion with respect to a point. Thus investigating nonlinear phenomena (e.g. symmetry breaking, merging crisis, coexisting attractors, and hysteresis) based on the circuit's symmetry is of great scientific relevance. Moreover, taking into account that a perfect symmetry is not realistic, the study of chaotic symmetric circuits in the presence of small symmetry breaking represents an interesting issue. In fact, the asymmetry deliberately induced in a nonlinear circuit can be controlled to reveal number of complex and interesting features (e.g. mixed mode oscillations, bursting, hysteresis, coexisting multiple asymmetric attractors, just to name a few) as in the case of Chua's circuit [8, 9], Helmtz-Duffing oscillator [10], parametrically excited pendulum [11-13], model of thermal convection [14], network of nonlinear circuits [15], Josephson equation [16], modified Coullet system[17], and memristive jerk circuits [18].

In this work, we consider the dynamics of the well-known Shinriki's oscillator both in its symmetric and asymmetric modes of operation. First of all, let us briefly recall the interesting works related to the Shinriki's oscillator [3-4] which is also regarded as a modified van der Pol oscillator. In Ref. [3], the authors showed that the Shinriki's circuit can produce oscillations with a random like waveform or a periodic oscillation depending on the selection of its parameters. Ref. [4] further investigates the dynamics of the Shinriki's oscillator from a qualitative viewpoint. The authors show that the circuit can develop a rich pattern of dynamical behaviours including fixed points, period-n oscillations, chaotic attractors, and so on. The investigation proceeds to identify and classify all of them through a bifurcation analysis (e.g. pitchfork, Hopf bifurcation, period doubling bifurcations, and periodic windows). Theoretical studies, numerical analyses, and laboratory experiments were presented [4]. By replacing the positive conductance formed of antiparallel branches of semiconductor diodes with memristive diodes, the modified circuit is shown to experience the coexistence of multiple periodic and chaotic attractors for the same parameter setting [19]. The above mentioned works highlight the particular interest devoted to the Shinriki's circuit and support the focus on the mathematical modelling and the effects induced by symmetry breaking of the Shinriki's oscillator in this paper. In fact, beside the rich pattern of dynamic motions (e.g. period-doubling, intermittency, merging crisis, bi-stability) already discussed in the original Shinriki's oscillator [4], the symmetry breaking induces new phenomena such coexisting bifurcation branches, coexisting asymmetric attractors, and hysteresis. The main contribution of this work lies in the following three key points:

(1) Instead of following a black box approach [4] leading to an approximate cubic-like current voltage characteristics of the nonlinear positive conductance of the Shinriki's circuit, we propose a close form model for the latter part by exploiting the Shockley diode equation. The result is a close form mathematical model of the whole Shinriki's oscillator that takes into account the intrinsic characteristics of semiconductor diodes forming the nonlinear part (i.e. the positive conductance).
(2) The above mentioned approach enables to address the realistic issue of symmetry breaking of the oscillator by considering different numbers of diodes within the two branches forming the positive conductance. The symmetry breaking analysis yields two asymmetric coexisting bifurcation branches (i.e. asymmetric bi-stability) each of which undergoes its own sequence of bifurcations to chaos when monitoring the main control parameter. In this special mode, the merging process never occurs; instead, one of the bifurcation branches vanishes when decreasing the control parameter beyond a critical value following a crisis event.
(3) The theoretical results are validated by carrying out laboratory experimental studies of the physical circuit.

The remainder of paper is structured as follows. Sect. 2 describes the modelling process of the Shinriki's oscillator under consideration. The circuit diagram of the oscillator is presented and a convenient mathematical model based on the Shockley diode model is proposed to investigate the complex dynamics of the oscillator. Sect. 3 discusses the number and nature of fixed points, and the symmetry properties of the model depending on the values of control parameters. The occurrence of Hopf bifurcation is also examined. Sect. 4 deals with the numerical study of the system. Several bifurcation diagrams and related plots of numerically computed largest Lyapunov exponents are plotted to illustrate various transitions to chaos. The phenomenon of asymmetric bi-stability is also revealed and illustrated by using phase portraits and basins of attractions as well. Laboratory experimental study of the oscillator represents the material of Sect. 5. Experimental measurements are in very good agreement with the theoretical study. Sect. 6 summarizes and concludes the whole work.

2. Circuit description and state equations

2.1. Circuit description

The circuit diagram of the Shinriki's oscillator [4] under consideration is provided in Fig. 1. This oscillator is an autonomous chaotic circuit belonging to the Chua's circuit family. It consists of a linear RLC circuit coupled to a linear RC oscillator through a nonlinear element, the positive conductance which is implemented here by using a set of semiconductor diodes. The power of the circuit is provided by negative impedance converter (NIC) realized with two identical resistors (R_1 and R_2), a linear operational amplifier (TL082CN), and a grounded resistor (R_3). The NIC operates as a linear negative resistor having the value $Z_N = -R_3$. The amount of current into the oscillatory part of the oscillator is adjusted via R_5. This tuneable resistor will be considered as the main bifurcation control element for the whole system. With the hypothesis of ideal op. amplifiers operating linearly, it is important to quote that the positive conductance constitute the only nonlinear component that accounts for the extremely complex and chaotic dynamics experienced by the whole electronic oscillator. In this work, the symmetry of the circuit is broken by constructing a positive conductance with six diodes in one branch and m $(1 \leq m \leq 5)$ diodes in the opposite branch (see Fig. 1). This strategy of symmetry breaking seems simpler that the common approach consisting of forcing a DC voltage source [20]. Moreover, it takes into account the real situation of parameters dispersion among the semiconductor diodes forming the positive conductance (i.e. non identical diodes).

2.2 State equations

The first step in the derivation process of the state equation of the oscillator in Fig. 1 is the mathematical description of the current voltage characteristics of the nonlinear positive conductance. By exploiting the Shockley exponential diode equation, the following expression can be easily obtained:

$$I_D = I_S\left(\exp\left(V_D/mnV_T\right) - \exp\left(-V_D/6nV_T\right)\right) \tag{1}$$

where I_S, n, V_T denote the reverse current, the emission coefficient, the thermal voltage of the diode respectively [21-22], and m, the number of diodes in the lower branch of the positive conductance (see Fig. 1). All over this work, we assume identical general purpose $1N4148$ semiconductor diodes which parameters are: $I_S = 2.682nA$, $n = 1.9$, $V_T = 26mV$. Denoting by I_L the current flowing through the inductor L, V_C, V_{C_0} the voltage across capacitors C, and C_0 respectively, the application of Kirchhoff's laws to the circuit diagram of Fig. 1 leads to the following set of three coupled first order differential equations (ODEs) governing the dynamics of the Shinriki's oscillator:

$$\begin{cases} C_0 \dfrac{dV_{C_0}}{dt} = \left(\dfrac{1}{R_3} - \dfrac{1}{R_4}\right) V_{C_0} + I_D\left(V_C, V_{C_0}\right) \\ C \dfrac{dV_C}{dt} = -I_L - \dfrac{V_C}{R_5} - I_D\left(V_C, V_{C_0}\right) \\ L \dfrac{dI_L}{dt} = V_C \end{cases} \tag{2a}$$

$$I_D = I_S \exp\left(\left(V_C - V_{C_0}\right)/mnV_T\right) - I_S \exp\left(-\left(V_C - V_{C_0}\right)/6nV_T\right) \tag{2b}$$

In order to ease both analytical and numerical analyses, the following change of variables and parameters (among many others [23, 24]) is convenient:

$$\rho = \sqrt{L/C},\ t = \tau\sqrt{LC},\ V_{ref} = 6nV_T,\ x_1 = V_C/V_{ref},\ x_2 = V_{C_0}/V_{ref},\ x_3 = \rho I_L/V_{ref},\ \eta = C/C_0 \\ ,\alpha = \rho/R_3 - \rho/R_4,\ \delta = \rho/R_5,\ k = 6/m \tag{3}$$

With this change of variables and parameters, the non-dimensional circuit model takes the form of a smooth nonlinear third order system:

$$\begin{cases} \dot{x}_1 = -x_3 - \delta x_1 - \gamma f_k\left(x_1 - x_2\right) \\ \dot{x}_2 = \eta\left(\alpha x_2 + \gamma f_k\left(x_1 - x_2\right)\right) \\ \dot{x}_3 = x_1 \end{cases} \tag{4a}$$

$$f_k(x) = \frac{1}{2}\left(\exp(kx) - \exp(-x)\right) \tag{4b}$$

where the over dot indicates derivation with respect to the non-dimensional time τ. We notice that only a couple of state variables (i.e. x_1 and x_2) contributes in the argument of the nonlinearity $f_k(.)$ of the model. This nonlinearity reduces to the hyperbolic sine function for the special case $k = 1.0$ corresponding to the symmetric mode of operation. Graphical representations of $f_k(.)$ for several discrete values of parameter k are provided in Fig. 2. Another interesting and important remark is that each of the three state variables is real and may be captured experimentally by using a classical oscilloscope. Five independent parameters can be identified in the evolution equations (4); however, unless otherwise mentioned, δ (i.e. the tuneable resistor R_5) will be considered all over this work, as the main bifurcation control parameter. The numerical experiment utilizes the following fixed values of electronic components: $L = 225mH$, $C_0 = 10nF$, $C = 100nF$, $R_1 = 5.6k\Omega$, $R_2 = 5.6k\Omega$, $R_3 = 10k\Omega$, $R_4 = 50k\Omega$, $R_5 - tuneable$, $D = 1N4148$ ($n = 1.9$, $V_T = 26mV$ and $I_S = 2.682nA$). With the adoption of this special set of components values, we obtain the following dimensionless parameters set: $\eta = 10.0$, $\alpha = 0.120$, $\sigma = 1.50$ and $\gamma = 8.143724 \times 10^{-5}$.

3. Basic properties of the model

3.1. Symmetry

In general for $k \neq 1.0$, system (4) is non-symmetric suggesting the absence of non-symmetric attractors. However, for the special case $k = 1.0$, it easy to see that system (4) is invariant under the transformation: $(x_1(\tau), x_2(\tau), x_3(\tau)) \Leftrightarrow (-x_1(\tau), -x_2(\tau), -x_3(\tau))$. Accordingly, if $(x_1(\tau), x_2(\tau), x_3(\tau))$ is a solution of system (4) for a fixed set of parameters values, then $(-x_1(\tau), -x_2(\tau), -x_3(\tau))$ is also a solution with the same set of parameters values. The origin $E_0(0,0,0)$ of the system coordinate satisfies the evolution equations (4) and thus represents a trivial equilibrium point. Consequently, any attractor of system (4) has to be symmetric with respect to the origin; otherwise it must possess a twin partner to fulfill the symmetry of the model. One of the practical implications of this symmetry is the presence of several coexisting symmetric attractors for several ranges of system parameters.

3.2 Equilibrium point analysis

The equilibrium points [25-30] of system (4) are found by setting to zero the derivatives and solving the corresponding nonlinear system. The following nonlinear and transcendental system is obtained:

$$\begin{cases} -x_3 - \delta x_1 - \gamma f_k(x_1 - x_2) = 0 \\ \eta(\alpha x_2 + \gamma f_k(x_1 - x_2)) = 0 \\ x_1 = 0 \end{cases} \tag{5}$$

Performing simple algebraic manipulations of system (5) yields the following simplified form:

$$\begin{cases} x_3 = \alpha x_2 \\ \alpha x_2 + \gamma f_k(-x_2) = 0 \\ x_1 = 0 \end{cases} \tag{6}$$

Obviously (6) represents a transcendental nonlinear system for which analytical solutions are not possible. The nontrivial fixed points however, can be found by exploiting numerical methods, in particular, by using the MATLAB built-in function "fzero" for any fixed values of system parameters. As already pointed out, the origin $E_0(0,0,0)$ represents the only trivial fixed point. Notice that the non-trivial fixed points (E_1 , E_2) can be obtained by taking the zeros of the nonlinear function $\psi_k(x)=\alpha x+\gamma f_k(-x)$ whose graphical representations are provided in Fig. 2 for several discrete values of parameter k. Notice that these fixed points are defined uniquely by the three parameters α , k and γ. Moreover, there are always two fixed independently of the value of the symmetry parameter k in addition to a third one whose location changes with k. Considering the numerical values of parameters mentioned above, the fixed points corresponding to two different values (i.e. $k=1.0$ and $k=1.5$) of the symmetry parameter k have been obtained as in Table 1 and 2. Considering a small perturbation of system (5) around any of the point $P\left(x_1^0,x_2^0,x_3^0\right)$, we derive the following Jacobian matrix :

$$M_J=\begin{bmatrix} -\delta-\gamma\vartheta & \gamma\vartheta & -1 \\ \eta\gamma\vartheta & \eta(\alpha-\gamma\vartheta) & 0 \\ 1 & 0 & 0 \end{bmatrix} \tag{7}$$

where $\vartheta=0.5k\exp\left(k\left(x_1^0-x_2^0\right)\right)+0.5\exp\left(-\left(x_1^0-x_2^0\right)\right)$. Thus, we examine the stability of each equilibrium point by solving the characteristic equation $\det(M_J-\lambda I_3)=0$ where I_3 denotes the 3x3 unity matrix [25-30]. Tables-1 and 2 show the roots of the characteristic equation computed for two discrete values of the symmetry parameter (i.e. $k=1.0$ and $k=1.5$) based on the Newton-Raphson method when monitoring parameter δ in the interval $0.10\le\delta\le0.15$. It clearly appears from Table-1 that for the critical value $\delta=\delta_H\approx0.123$, there is a real root and a pair of complex conjugate eigenvalues with zero real part related to the pair of nontrivial symmetric fixed points (E_1,E_2). This critical value δ_H denotes the Hopf bifurcation point in the symmetric mode of oscillation. In fact, for values of $\delta\ge\delta_H$, the symmetric par of equilibrium points (E_1,E_2) has eigenvalues all with negative real part typical of stable spiral points. Eigenvalues having positive real parts are found by considering values of $\delta\le\delta_H$. This situation is typical of unstable spiral points. Obviously, the origin $E_0(0,0,0)$ is always unstable both in the symmetric and asymmetric modes of oscillations provided that the corresponding eigenvalues have real part of different signs. Similar conclusions are drawn from Table 2 when examining the evolution of fixed points and corresponding eigenvalues for varying δ . From the above analysis, it follows that the system experiences self-excited attractors [31-35] both in the symmetric and asymmetric modes of operation.

4. Numerical study

In this part, we are interested on the chaos mechanism in the Shinriki's oscillator both in its symmetric and asymmetric modes of operation. To this end, system (4) is solved numerically using the most common fourth-order Runge-Kutta integration scheme. We adopt a constant time grid $\Delta t = 0.005$ with variables and parameters in extended precision mode all over the numerical analysis. We illustrate the scenario to chaos by using graphical nonlinear analysis tools such as phase space trajectory plots, time series, graphs of maximal Lyapunov exponents (MLE), bifurcation diagrams, and Poincare section as well. The Lyapunov exponents are computed following the well-known Wolf algorithm [36]. We show that the symmetry breaking greatly influences the dynamics of the Shinriki's oscillator.

4.1 Scenario to chaos in the symmetric system

As in the reference paper by Freire and colleagues, parameter δ (i.e. resistor R_5) is considered as main control parameter; the rest of parameters having the numerical values assigned in Sect.2. The numerical experiment shows that for values of $\delta > \delta_H \approx 0.123$, the system develops a fixed point motion. If the control parameter is decreased beyond the critical value δ_H, the system undergoes a Hopf bifurcation yielding a pair of coexisting symmetric period-1 cycles. Further decreasing δ, the system experiences a period-doubling route to chaos resulting to a pair of mutually symmetric spiraling strange attractors. The merging crisis event occurs at $\delta = \delta_{cr1} \approx 0.065$ and a double-band strange attractor is obtained. The above sequence of bifurcations is perfectly illustrated by the bifurcation diagram of Fig.3 (upper panel). The plot of maximal Lyapunov exponent versus δ corresponding to the bifurcation diagram of Fig.3 is shown in the lower panel. From this latter figure, bands of chaos marked by a positive Lyapunov exponent can be easily distinguished from those of regular motion with negative values of the same exponent. In the graph of Fig. 3, blue and magenta data are obtained by scanning the parameter downward (i.e. downward continuation) starting from different initial conditions $x_1(0) = 0, x_2(0)$, $x_3(0) = \pm 1$. Fig.4 shows examples of numerically computed phase portraits for some discrete values of parameter δ confirming the scenario to chaos described above. Moreover the complexity of the attractor in Fig.4f is better illustrated by the corresponding frequency spectrum, Poincaré section, the 2D projections, and 3D View provided in Fig. 5. Notice that the Poincaré section (see Fig. 5c) consists of a set of points scattered in the state space, and the related signal depicts a broadband noise like frequency spectrum (see Fig. 5d) typical of chaotic attractors.

4.2 Effects of symmetry breaking

The bifurcation diagram of the system (4) and corresponding graph of MLE showing the system response in the presence of symmetric breaking are presented in Fig.6. Blue and magenta colors correspond to downward scanning of parameter δ as mentioned previously. Both coexisting bifurcation branches depict completely different bifurcation sequences typical of bi-stable systems. It is interesting to quote that both bifurcation branches emanate from Hopf bifurcation related to the pair of asymmetric fixed points E_1 and E_2. Interestingly, the magenta branch vanishes at the critical value $\delta_{c2} \approx 0.063$ giving room to a mono stable behavior for any value of δ in the range $0.1 \leq \delta \leq \delta_{cr2}$. The bi-stable response of the system with respect to parameter δ can also be captured from sample phase portraits of Fig.7 produced for some discrete value of δ. The presence of two different attractors magnetizes the space as illustrated in Fig.8 (a, b). The fact that the basin of attraction is nearly the entire phase space in Fig.8c confirms the presence of a single attractor for the specific parameter value. We further illustrate the impact of symmetry breaking on the system response by providing in Fig. 9 several bifurcation diagrams of the system for other discrete values of parameter k. As already pointed out one the bifurcation branch always vanishes when decreasing the control parameter δ giving room to a mono stable behavior.

5. Experimental Study

5.1 Circuit set-up

We now focus on the experimental study [37-39] of the Shinriki's oscillator in order to verify the extremely complex and striking behaviors predicted by the theoretical analysis presented above. For the purpose, the circuit diagram in Fig. 1 is realized on a breadboard. The experimental oscillator uses TL082 op. amplifiers integrated circuits type powered by a symmetric $\pm 12V$ dc voltage source. Inductor L is constructed following gyrators technique [39]. We keep the same values of electronics components employed for the numerical experiment in order to facilitate the comparison process. Both symmetric and asymmetric modes of operation are considered. In particular, we use 6 diodes in both branches of the positive conductance in the symmetric mode while both branches contain different number of diodes (e.g. 6 and 4) in the asymmetric mode. We obtain the experimental results by decreasing gradually R_5 and display phase-space trajectories $\left(V_C, V_{C_0}\right)$ using a double channel oscilloscope in the XY mode. Experimental phase portraits are captured by using a high resolution digital camera.

5.2 Behaviour of the symmetric circuit

Considering the symmetric setup of the oscillator, we note that when the control resistor R_5 is slowly increased it is found that the experimental oscillator experiences the familiar period doubling scenario to chaos starting from the Hopf bifurcation. For $R_5 = 17.11 k\Omega$ a pair of mutually symmetric period-1 limit cycles is observed. When R_5 is gradually increased, the whole sequence of bifurcation described above during the theoretical study is observed. In particular, the double band chaotic attractor is formed via merging crisis mechanism. Small windows of periodic oscillations occurring within the chaotic regions are also detected experimentally. A clear illustration of these observed phenomena can be captured from the experimental pictures of Fig. 10. Note that each mutually symmetric pair of coexisting attractors of Fig. 10 is obtained by switching on and off the power supply. A perfect similarity can be noticed between numerical and experimental phase portraits which may serve to validate the mathematical model proposed in this work.

5.3 Behaviour of the asymmetric circuit

In the asymmetric mode of operation, the circuit develops a fixed point motion and no oscillations are observed if the control resistor R_5 remains lower the critical value $R_{5H} \approx 10k\Omega$. If the control resistor is increased past this critical value R_{5H}, the system undergoes a Hopf bifurcation giving rise to two coexisting asymmetric limit cycles of different size (see Fig. 11). By further increasing the control resistor R_5 without resetting the circuit, each of the limit cycles follows its own sequence of bifurcations to chaos. Interestingly, the sequence of bifurcations related to the attractor with smaller size vanishes at the critical value $R_{5C} \approx 22.435k\Omega$ giving room to a mono-stable circuit (see Fig. 11). For any value of R_5 in the range $R_{5H} \leq R_5 \leq R_{5C}$ coexisting attractors are observed by turning of and off the power supply. In contrast to the symmetric mode of operation, no merging process occurs in the asymmetric setup. As already mentioned above, a very good similarity can be observed between numerical and experimental phase portraits (see Fig. 5 and 11). However, a slight shift in the values of the control resistor R_5 can be noted between numerical and experimental results. This discrepancy can be attributed to the precision on the values of electronic components and the simplifying assumptions adopted during the mathematical modeling of the circuit (i.e. ideal diode model, ideal op. amplifier).

6. Concluding remarks

This paper investigated the dynamics of the well-known Shinriki's oscillator oscillator based on a new mathematical model that takes into account the intrinsic properties of semiconductor diodes forming the positive conductance of the underlined oscillator. Both symmetric and asymmetric modes of operation were considered. The symmetry breaking is achieved by considering different number of semiconductor diodes in the two branches forming the nonlinear part (i.e. the positive conductance). Using standard nonlinear analysis techniques such as bifurcation diagrams, phase portraits, graphs of Lyapunov exponents' plots, time series, and basins of attractions, the dynamics of the system has been categorized with respect to its control parameters. In the symmetric mode of operation, the bifurcation analysis indicates that chaos arises in the Shinriki's oscillator following period-doubling and symmetry restoring crises phenomena when monitoring the control parameter in small steps. Of particular interest, it is found that the Shinriki's oscillator demonstrates two asymmetric coexisting bifurcation branches (i.e. asymmetric bi-stability) each of which exhibits its own sequence of bifurcations to chaos when monitoring the main control parameter. In this special mode, the merging process never occurs; instead, one of the bifurcation branches suddenly collapses when decreasing the control parameter beyond a critical value following a crisis event. The theoretical results are confirmed by carrying out laboratory experimental measurements on a physical prototype. As far as the authors' knowledge goes, these interesting features have not yet been discussed in the Shinriki's circuit and thus deserve dissemination.

An interesting issue under consideration is that of proposing several synchronization strategies for the Shinriki's oscillator considered in this paper owing to potential technological exploitations (e.g. Chaos communication, image coding, random signal generation and so on) and academic applications as well.

References

[1] Pivka L, Wu CW, Huang A (1994) Chua's oscillator: A compendium of chaotic phenomena. Journal of the Franklin institute 331B (6), 705-741

[2] Fotsin HB, Woafo P (2005) Adaptive synchronization of a modified and uncertain chaotic Van der Pol-Duffing oscillator based on parameter identification. Chaos Solitons and Fractals 24; 1363-1371

[3] Shinriki M, Yamato M, Mori S (1981) Multimode oscillations in a modified van der Pol oscillator containing a positive nonlinear conductance. Proc. IEEE 69, 394-395.

[4] Freire E, Franquelo LG, Aracil J (1984) Periodicity and chaos in an autonomous electrical system. IEEE Trans. Circuits Syst. Vol. Cas-31 (3), 237-247

[5] Kengne J, Njitacke ZT, Nguomkam NA, Fouodji Tsotsop M, Fotsin HB (2015) Coexistence of multiple attractors and crisis route to chaos in a novel chaotic jerk circuit. Int. J. Bifurcation and Chaos 25 (4): 1550052

[6] Kengne J, Nguomkam NA, Njitacke Z T (2017) Antimonotonicity, Chaos and Multiple Attractors in a Novel Autonomous Jerk Circuit. I. J. Bifurcation and Chaos 27(7): 1-20

[7] Kengne J, Kenmogne F (2014) On the modeling and nonlinear dynamics of autonomous Silva-Young type chaotic oscillators with flat power spectrum. Chaos 24: 043134.

[8] Kahllert C (1993) The effects of symmetry breaking in Chua's circuit and related piecewise-linear dynamical systems. Int. J. of Bif. and chaos ; 3(4): 963-979

[9] Dana SK, Chakraborty S, Ananthakrishna G (2005) Homoclinic bifurcation in Chua's circuit. Pramana journal of Physics; 64(3):44344

[10] Cao H, Seoane JM, Sanjuan, MAF (2007) Symmetry-breaking analysis for the general Helmholz-Duffing oscillator. Chaos Solitons and fractals; 34: 197-212

[11] Zhou P, Cao H (2007): The effects of symmetry breaking on the parameterically excited pendulum, Chaos Solitons and fractals; 38: 590-597

[12] Sofroniou A, Bishop SR (2006) Breaking the symmetry of the parametrically excited pendulum. Chaos Solitons and Fractals; 28:673-81.

[13] Bishop SR, Sofroniou A, Shi P (2005) Symmetry-breaking in the response of the parameterically excited pendulum model. Chaos Solitons and Fractals; 25(2): 27-264

[14] Rynio R, Okninski A (1998) Symmetry breaking and Fractal dependence on initial conditions in dynamical systems: Ordinary differential equations of thermal convection. Chaos Solitons and Fractals; 9(10): 1723-1732

[15] Henrich M, Dahms T, Flunkert V, Teitsworth SW, Scholl E (2010) Symmetry breaking transitions in networks of nonlinear circuits elements. New journal of physics; 12:113030

[16] Cao H, Jing Z (2001) Chaotic dynamics of Josephson equation driven by constant and ac forcings. Chaos Solitons and Fractals; 12: 1887-1895

[17] Kengne J, Kengne LK (2018) Scenario to chaos and multistability in a modified Coullet system: effects of broken symmetry. Int. J. Dynam. Control DOI:10.1007/s40435-018-0483-2

[18] Hua, M., Yang, S., Xu, Q., Chen, M., Wu, H., & Bao, B. (2020). Forward and reverse asymmetric memristor-based jerk circuits. *AEU-International Journal of Electronics and Communications*, 153294.

[19] Kengne J, Njitacke ZT, Kamdoum TV, Nguomkam NA (2015) Periodicity chaos and multiple attractors in a memristor-based Shinriki's oscillator. Chaos 25(10):103126

[20] Liu W, Guo Y, Lu M, Liu C (2010) Chaos attractor compressing and expanding in Chua circuit driven by a direct current voltage. In Proc. Of the third Int. Symposium on information science and engineering. DOI: 10.1109/ISIE.2010.105; 446-450

[21] Hanias MP, Giannaris G, Spyridakis AR (2006) Time series analysis in chaotic diode resonator circuit". Chaos, Solitons and Fractals; **27**:569.

[22] Sukov DW, Bleich ME, Gauthier J, Socolar JES (1997) Controlling chaos in a fast diode resonator using extended time-delay auto-synchronization: Experimental observations and theoretical analysis. Chaos; **7**(4), 560-576.

[23] Wu, H., Ye, Y., Chen, M., Xu, Q., & Bao, B. (2019). Extremely slow passages in low-pass filter-based memristive oscillator. *Nonlinear Dynamics*, *97*(4), 2339-2353.

[24] Bao, B. C., Wu, P. Y., Bao, H., Xu, Q., & Chen, M. (2018). Numerical and experimental confirmations of quasi-periodic behavior and chaotic bursting in third-order autonomous memristive oscillator. *Chaos, Solitons & Fractals*, *106*, 161-170.

[25] Strogatz SH (1994) *Nonlinear dynamics and chaos*, Reading, Addison-Wesley;

[26] Jiang, X., Chen, X., Huang, T., & Yan, H. (2020). Bifurcation and Control for a Predator-Prey System with Two Delays. *IEEE Transactions on Circuits and Systems II: Express Briefs*.

[27] Nayfeh AH, Balachandran B (1995) *Applied nonlinear dynamics: Analytical, computational and experimental methods*, John Wiley & Sons, New York;

[28] Kuznetsov YA (1995) *Elements of Applied Bifurcation Theory*. New York: Springer-Verlag;

[29] Jiang, X., Chen, X., Chi, M., & Chen, J. (2020). On Hopf bifurcation and control for a delay systems. *Applied Mathematics and Computation*, *370*, 124906.

[30] Jiang, X. W., Zhan, X. S., Guan, Z. H., Zhang, X. H., & Yu, L. (2015). Neimark–Sacker bifurcation analysis on a numerical discretization of Gause-type predator–prey model with delay. *Journal of the Franklin Institute*, *352*(1), 1-15.

[31] Kuznetsov NV, Leonov GA, Yuldashev MV, Yuldashev RV (2017) Hidden attractors in dynamical models of phase-locked loop circuits: limitations of simulation in MATLAB and SPICE, Communications in Nonlinear Science and Numerical Simulation; **51**: 39-49

[32] Leonov G, Kuznetsov N, Vagaitsev V (2012) Hidden attractor in smooth Chua systems. Physica D: Nonlinear Phenomena; 241(18): 1482-1486

[33] Leonov GA, Kuznetsov NV, Mokaev TN (2015) Homoclinic orbits, and self-excited and hidden attractors in a Lorenz-like system describing convective fluid motion. European Physical Journal Special Topics; 224: 1421–1458

[34] Jafari A, Mliki E, Akgul A, Pham VT, Kingni ST, Wang X, Jafari S (2017) Chameleon : the most hidden chaotic flow. Nonlinear Dyn; DOI 10.1007/s11071-017-3378-4.

[35] Jafari S, Sportt JC, Nazarimehr F (2015) Recent new examples of hidden attractors. Eur. Phys. J. Spec. Top; 224: 1469–1476

[36] Wolf A, Swift JB, Swinney HL, Wastano JA (1985) Determining Lyapunov exponents from time series. Physica D; **16**: 285-317

[37] Wang YF, Qiao ST, Jiang JG (2008) Simulation of chaos in asymmetric nonlinear Chua's circuit. Shanghai Jiaotong University (Sci), 13(4) 453-456

[38] Healey JJ, Broomhead DS, Cliffe KA, Jones R, Mulin T (1991) The origin of chaos in a modified Van der Pol oscillator. Pysica D 4:322-339

[39]Makoto I (2001) Synthesis of electronic circuits for simulating nonlinear dynamics, *Int. J. of Bifurcation and Chaos,* **11**, 605-653

LIST OF TABLES

Table 1: Eigenvalues and stability of fixed points computed for some discrete values of the bifurcation control parameter δ in the symmetric mode of oscillations ($k = 1.0$).

Values of the Bifurcation parameter (δ)	Eigenvalues $(\lambda_1, \lambda_2, \lambda_3)$ at nontrivial equilibria (E_1, E_2)	Eigenvalues $(\lambda_1, \lambda_2, \lambda_3)$ at the origin $E_0(0,0,0)$
$\delta = 0.1$	$0.0103 \pm 0.9452i$	$-0.0500 \pm 0.9987i$
	-14.1078	1.1997
	(unstable)	(unstable)
$\delta = 0.12$	$0.0013 \pm 0.9452i$	$-0.0600 \pm 0.9982i$
	-14.1099	1.1997
	(unstable)	(unstable)
$\delta_H = 0.123$	$0 \pm 0.9452i$	$-0.0615 \pm 0.9981i$
	-14.1102	1.1997
	(neutral, Hopf bifurcation)	(unstable)
$\delta = 0.125$	$-0.009 \pm 0.9452i$	$-0.0625 \pm 0.9980i$
	-14.1105	1.1997
	(stable)	(unstable)
$\delta = 0.15$	$-0.121 \pm 0.9450i$	$-0.0750 \pm 0.9972i$
	-14.1131	1.1997
	(stable)	(unstable)

Table 2: Eigenvalues and stability of fixed points computed for some discrete values of the bifurcation control parameter δ in an asymmetric mode of oscillations ($k = 1.5$).

Values of the Bifurcations parameter (δ)	Eigen values $(\lambda_1, \lambda_2, \lambda_3)$ at nontrivial equilibrium $E_1(0, 11.53, 1.3839)$	Eigenvalues $(\lambda_1, \lambda_2, \lambda_3)$ at nontrivial equilibrium $E_2(0, -7.392, -0.887)$	Eigenvalues $(\lambda_1, \lambda_2, \lambda_3)$ at the origin $E_0(0,0,0)$
$\delta = 0.1$	$0.0103 \pm 0.9452i$ -14.1078 (unstable)	$0.0103 \pm 0.9449i$ -13.563 (unstable)	$-0.0500 \pm 0.9987i$ 1.1997 (unstable)
$\delta = 0.12$	$0.0013 \pm 0.9452i$ -14.1099 (unstable)	$0.0014 \pm 0.9949i$ -13.5651 (unstable)	$-0.0600 \pm 0.9982i$ 1.1997 (unstable)
$\delta_H = 0.123$	$0.0 \pm 0.9452i$ -14.1102 (neutral, Hopf bifurcation)	$0.0 \pm 0.9449i$ -13.5655 (neutral, Hopf bifurcation)	$-0.0615 \pm 0.9981i$ 1.1997 (unstable)
$\delta = 0.125$	$-0.0009 \pm 0.9952i$ -14.1105 (stable)	$-0.0009 \pm 0.9449i$ -13.5657 (stable)	$-0.0625 \pm 0.9980i$ 1.1997 (unstable)
$\delta = 0.15$	$-0.0121 \pm 0.9950i$ -14.1131 (stable)	$-0.0120 \pm 0.9447i$ -13.5683 (stable)	$-0.0750 \pm 0.9972i$ 1.1997 (unstable)

LIST OF FIGURES

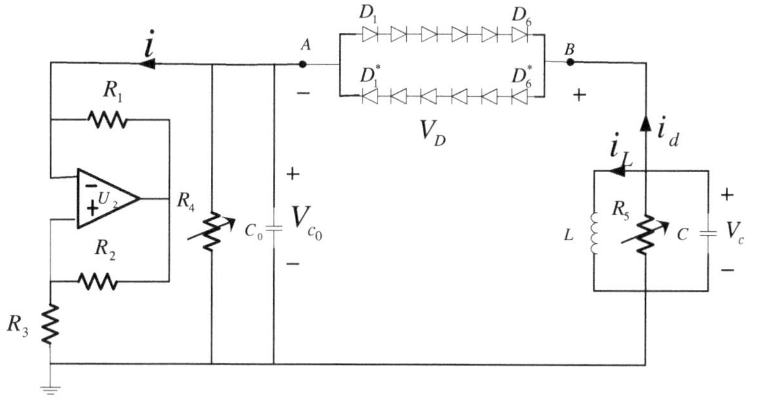

Fig. 1: Schematic diagram of the Shinriki-Yamamoto-Mori circuit in its symmetric mode of operation with 12 diodes forming the positive conductance. The op. amplifier and related resistors implement the NIC which value is given by $Z_{NIC} = -R_3$. The symmetry breaking is obtained by considering less than six (i.e. $1 \le m \le 5$) diodes in the lower branch of the positive conductance.

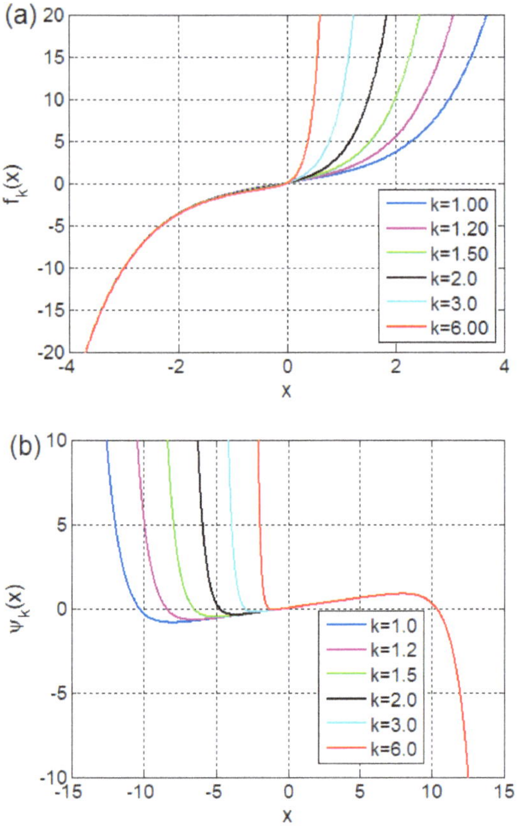

Fig. 2: Graphical reprentation of the nonlinear function $f_k(x)$ (a) for six discrete values of parameter k. The zeros of the nonlinear function $\psi_k(x)$ determine the equilibria of the system. Remark that there is always a pair of fixed zeros and a third one whose location is determined by parameter k.

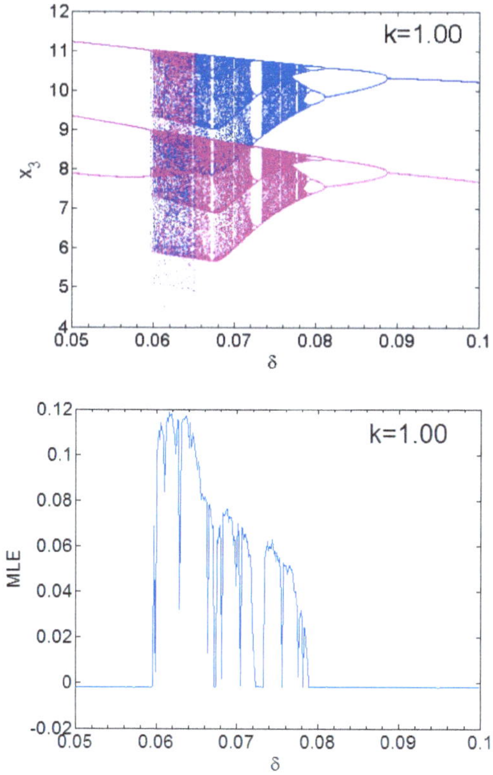

Fig. 3: (a) Bifurcation diagrams displaying local maxima of the coordinate x_3 versus parameter δ and the corresponding graph of maximum Lyapunov exponent (b) for the symmetric mode of oscillation. A positive value of MLE indicates chaos while regular oscillations are related to negative or null values of MLE.

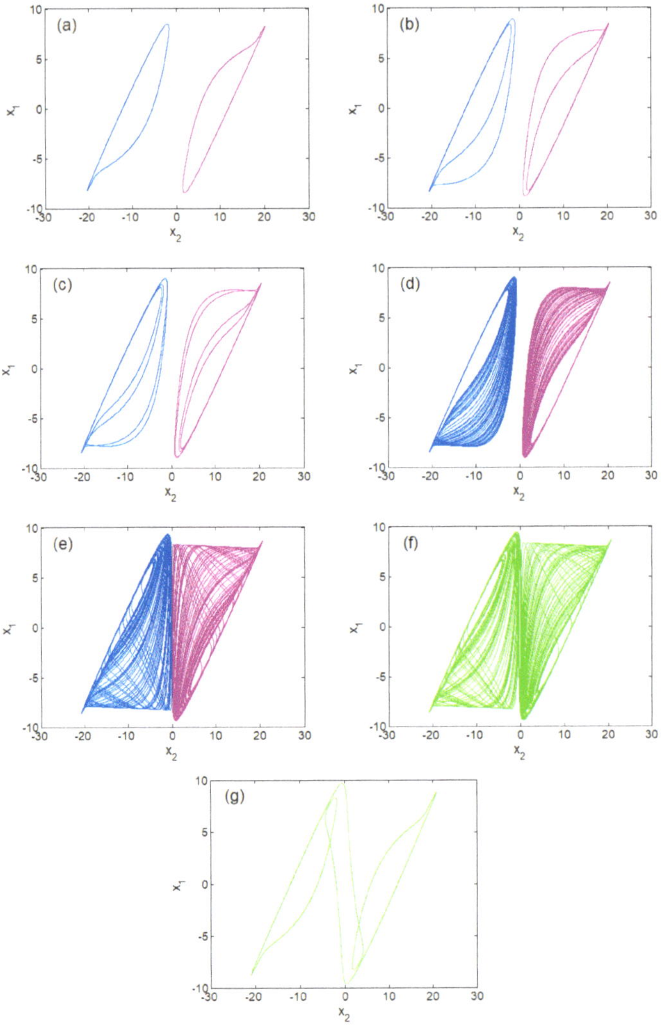

Fig. 4: Computer generated phase portraits of the system showing coexisting symmetric pairs of attractors for some discrete values of the control parameter δ: (a) two period-1 cycles for $\delta = 0.100$; (b) two period-2 cycles for $\delta = 0.084$; (c) two period-4 cycles for $\delta = 0.080$; (d) two chaotic attractors for $\delta = 0.076$; (e) two chaotic attractors for $\delta = 0.066$; (f) a double scroll chaotic attractor for $\delta = 0.0625$; (g) a symmetric period-3 cycle for $\delta = 0.055$. Initial conditions $(x_1(0), x_2(0), x_3(0))$ are $(0, 0, \pm 1)$.

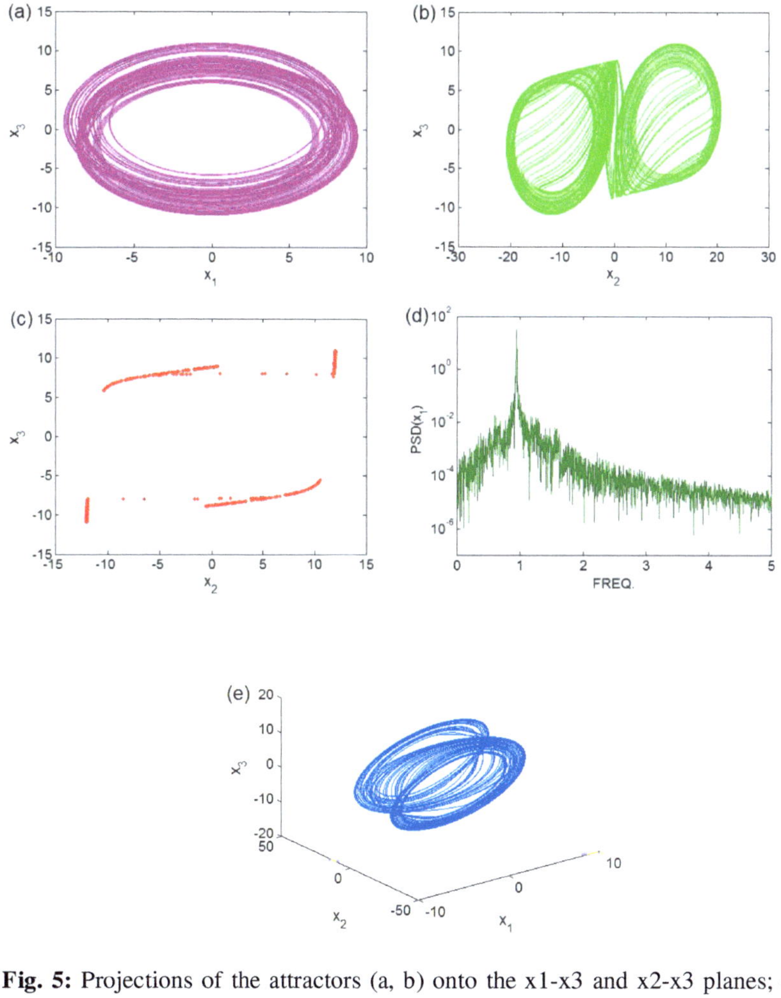

Fig. 5: Projections of the attractors (a, b) onto the x1-x3 and x2-x3 planes; double-sided Poincare section (c), frequency spectra (d), and 3D view of the double-scroll chaotic attractor computed for $k = 1.00$, $\delta = 0.0625$.

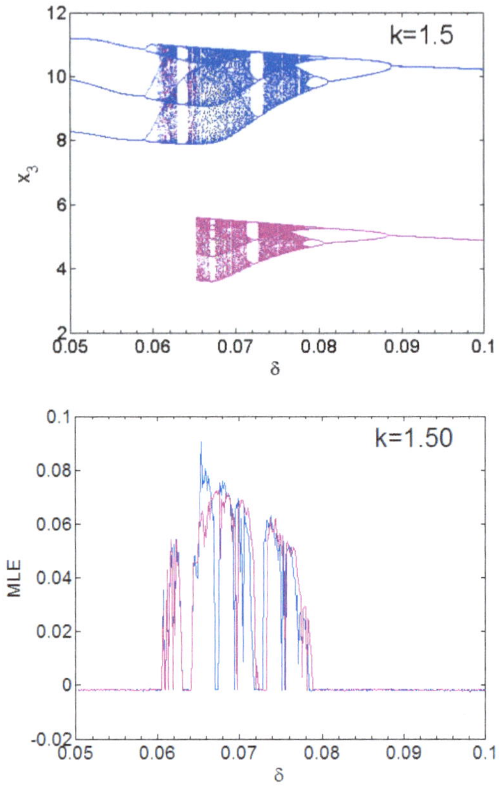

Fig. 6: Bifurcation diagrams displaying local maxima of the coordinate x_3 versus parameter δ and the corresponding graph of maximum Lyapunov exponent (MLE) computed in the asymmetric mode of oscillation with $k = 1.50$. Blue and magenta colors correspond to downward continuations of the control parameter starting with initial conditions $(0, 0, \pm 1)$ respectively.

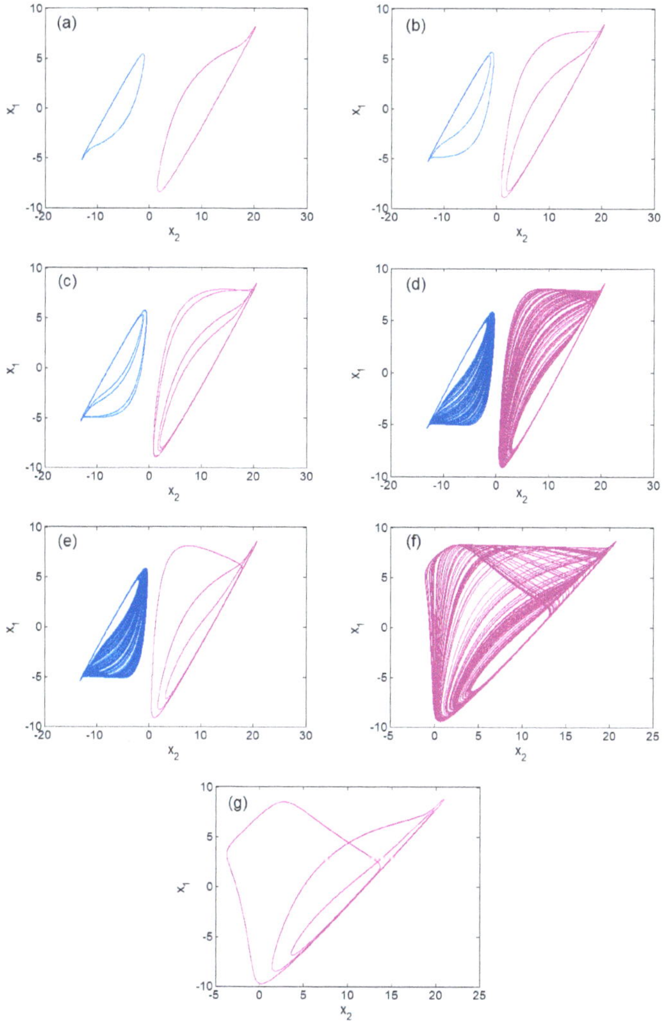

Fig. 7: Computer generated phase portraits of the system showing coexisting asymmetric modes of oscillations for some discrete values of the control parameter δ : (a) two period-1 cycles for $\delta = 0.100$; (b) two period-2 cycles for $\delta = 0.0825$; (c) two period-4 cycles for $\delta = 0.080$; (d) two chaotic attractors for $\delta = 0.074$; (e) a period-3 cycle and a chaotic attractor for $\delta = 0.073$; (f) a chaotic attractor for $\delta = 0.0625$; (g) a period-3 cycle for $\delta = 0.050$. Initial conditions $(x_1(0), x_2(0), x_3(0))$ are $(0, 0, \pm 1)$.

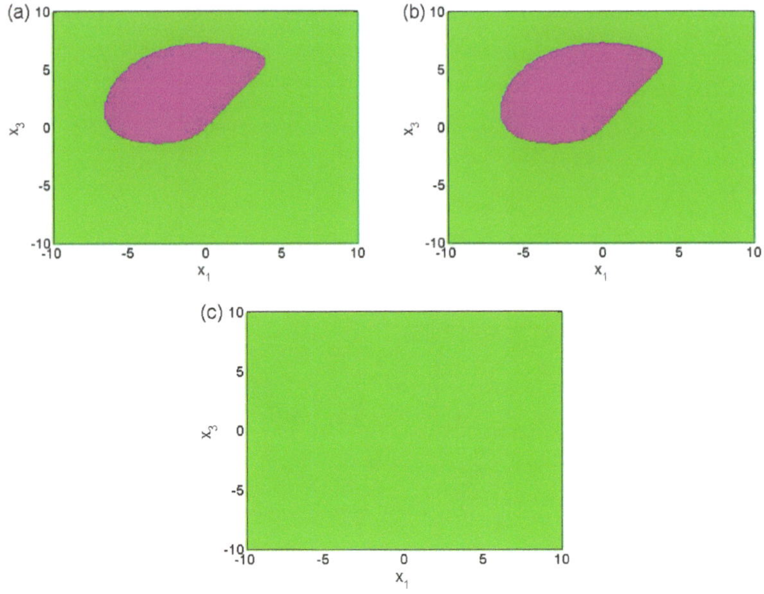

Fig. 8: Structure of the section (at $x_2 = 0$) of the basin of attraction for $k = 1.50$ computed for three discrete values of parameter δ : (a) $\delta = 0.074$, (b) $\delta = 0.0653$, and (c) $\delta = 0.0625$. Magenta zone correspond to the smaller attractor while green zones are related to the bigger one. Notice that in (c) there is a single attractor as the basin is nearly the whole state space.

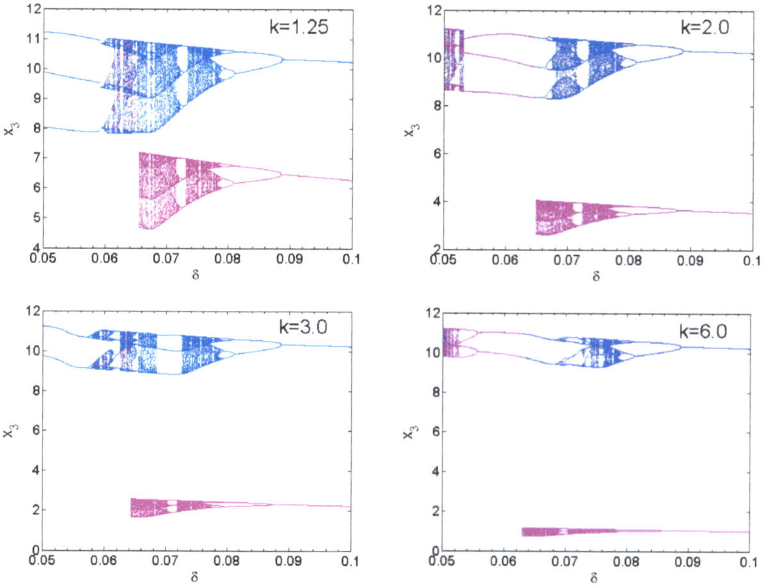

Fig. 9: Bifurcation diagrams displaying local maxima of the coordinate x_3 versus parameter δ computed for four discrete values ($k = 1.25$; 2.00 ; 3.00 ; 6.00) of asymmetry parameter k. We notice the strong impact of k on the dynamics of the system. In particular, the magenta branch disappears when decreasing the control parameter δ for any value of $k \neq 1.0$.

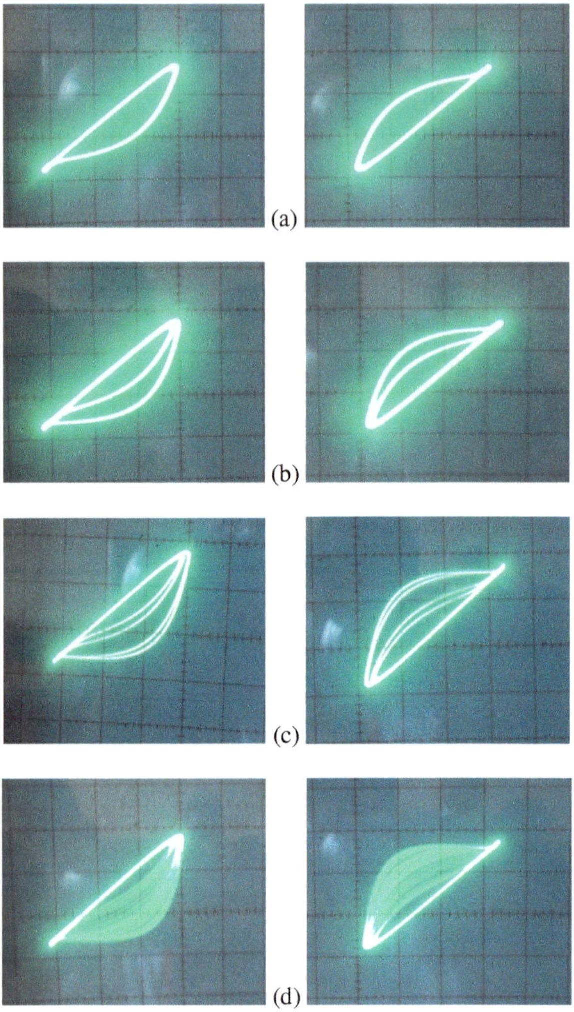

Fig. 10: Experimental phase portraits showing coexisting attractors and the period doubling scenario to chaos captured from the circuit: (a) period-1 cycles for $R_5 = 17.11k\Omega$; (b) period-2 cycles for $R_5 = 19.30k\Omega$; (c) period-4 cycles for $R_5 = 20.36k\Omega$; (d) spiraling chaos for $R_5 = 21.12k\Omega$; (e) screw-like chaos for $R_5 = 23.51k\Omega$; (f) double-scroll chaos $R_5 = 26.26k\Omega$; (g) period-3 cycle for $R_5 = 27.80k\Omega$. The scales are

$X = Y = 2V / div$. Coexisting mutually symmetric attractors are obtained by switching on and off the power supply for fixed values of the control resistor R_5.

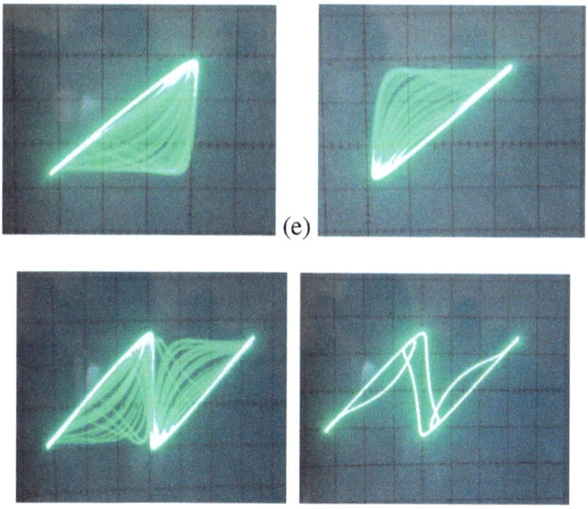

(e)

(f) (g)

Fig. 10 (continued)

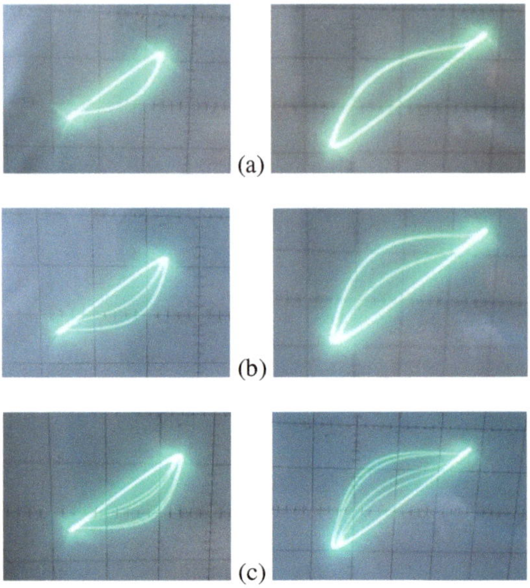

Fig. 11 : Experimental phase portraits showing coexisting attractors and the period doubling scenario to chaos captured from the circuit: (a) period-1 cycles for $R_5 = 16.85k\Omega$; (b) period-2 cycles for $R_5 = 17.31k\Omega$; (c) period-4 cycles for $R_5 = 17940k\Omega$; (d) spiraling chaos for $R_5 = 18.76k\Omega$; (e) coexistence of a chaotic attractor with a period-3 cycle for $R_5 = 19.26k\Omega$; (f) screw like chaos $R_5 = 22.67k\Omega$; (g) period-3 cycle for $R_5 = 28.20k\Omega$. The scales are $X = Y = 2V / div$. Coexisting mutually symmetric attractors are obtained by switching on and off the power supply for fixed values of the control resistor R_5.

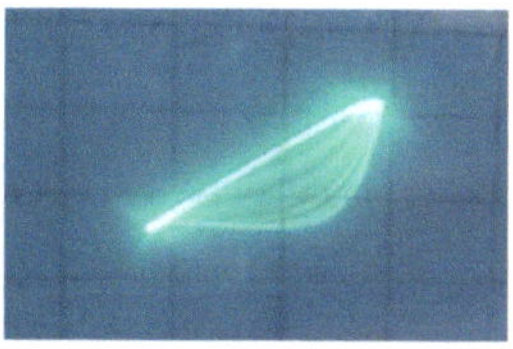

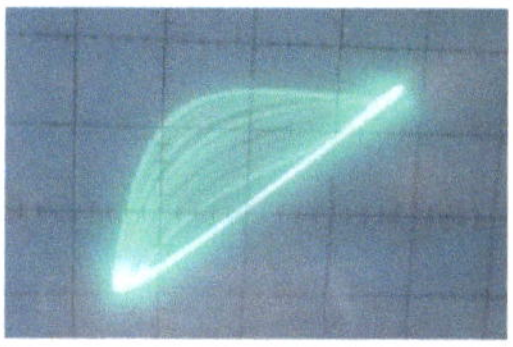

(d)

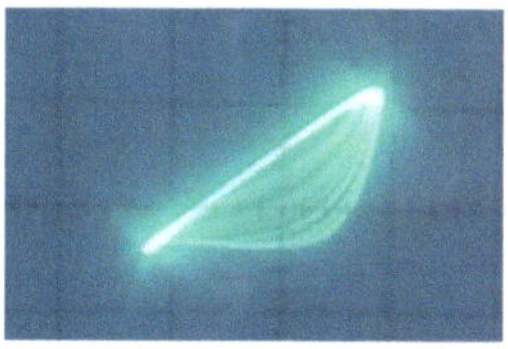

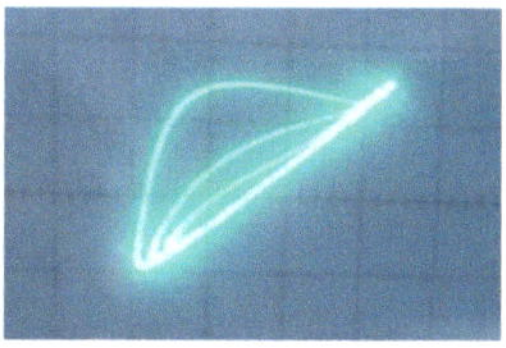

(e)

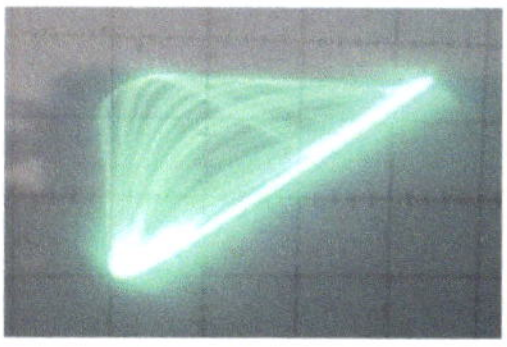

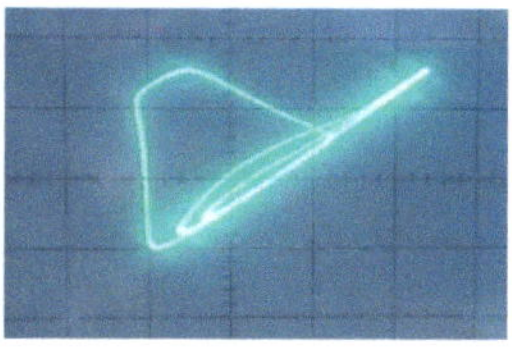

(f) (g)

Fig. 11 (continued)

www.ingramcontent.com/pod-product-compliance
Lightning Source LLC
LaVergne TN
LVHW071909071126
851857LV00033B/9
9781639028399